CRYSTAL VISIONING

a crystal workbook

Serenity Peterson

INTERDIMENSIONAL PUBLISHING
a division of
INTERDIMENSIONAL PRODUCTIONS
ISBN-0-932389-00-7

This book is dedicated to Daniel, Brooke, Billy, and all the beautiful Light Beings, seen and unseen, who helped me to channel this gift of Love.

DISCLAIMER

The healing crystals offered for sale by the author are not in themselves alleged to be a cure for any ailment, disorder or physical condition. Furthermore, it is not the intention of the author or publisher to prescribe for any conditions, physical or otherwise. The manner in which this material is interpreted and used is the sole responsibility of the reader.

CHAPTERS

INTRODUCTION .... *v*
THE MAGIC *1*
THE CALLING *4*
CHOOSING A CRYSTAL *7*
THE DEDICATION *14*
CLEARING TECHNIQUES *17*
ESOTERIC ANATOMY *24*
MEDITATION TECHNIQUES *30*
HEALING TECHNIQUES *36*
COMBINING OTHER GEM ENERGIES WITH CLEAR QUARTZ CRYSTAL *40*
POSTSCRIPT *61*

INTRODUCTION

I have left several blank pages at the back of this book for a very good reason — it is a workbook! It is my belief that we learn best by doing. No two experiences are exactly alike — that is the beauty of life! During your work with crystals, I hope that you will feel free to experiment. You may notice certain reactions within yourself, many revelations, that you will want to make note of. Do so on these pages. You, my dear friend, are my co-author!

1

THE MAGIC

Below the earth's surface lies an enchanted world filled with exquisite life forms—the crystal kingdom. Yet, strangely enough, many men, in their ignorance, have not seen the magical gifts these crystals hold. Instead, they have considered them to be simply inanimate, dead matter, to be used in the making of watches, microchips, lasers, optical lenses, or as an occasional ornament for their bodies or living spaces.

In actuality, these crystals are far from inanimate. They generate their own measurable electro-magnetic field. They are capable of holding a charge for long periods of time. They reproduce themselves, and will continue to do so, as well as to keep on growing, even when they are removed from the earth. All of these are facts which can be verified by any reputable scientist.

However, there are many more qualities inherent in quartz crystal. It is these qualities, which we have forgotten, but are now on the brink of rediscovering, which make them such friends and teachers of man on his evolutionary path toward acceptance of his Divinity.

They are a great tool for healing yourself or others. They are capable of concentrating and then transmitting healing energy—sometimes for long distances.

The crystals are great protectors, generating a field of almost impenetrable Divine energy, which is capable of transmuting any inappropriateness within their area.

They are exquisite channelers and transmitters of Divine energy, gently raising the vibrations of all within their vibratory field.

They form a connection between man and earth. In so doing, they re-establish the love affair between man and the planet. Through this Love, much healing occurs, not only for man, but for the planet as well. Balance will be restored on Earth.

Crystals also help man to see through the forms to realize that the same Divine energy is in everything, continually flowing, swirling, dancing throughout the Universe in ever-changing patterns. They enable him to look past the patterns to the energy beyond, and to use this energy for Divine Creation. For this reason, they are excellent meditation enhancers.

Crystals are also capable of absorbing and holding energy patterns for long periods of time. I know a beautiful couple who, on their wedding day, exchanged crystal pendants instead of rings. They wear them constantly. Whenever they are going to be away from each other for a time, they switch pendants, so that each will carry the other's essence with them wherever they are.

The very genesis and form of the crystals is a wonder in itself! The fact that such smooth, clear, perfectly faceted jewels could grow through heat and pressure in the damp mud of the earth is a miracle.

They have waited for so long to share what they have with you. Reach out and take what they have to offer you in Love: the knowledge of your connection with all life forms, the wisdom of your own Divinity!

2

THE CALLING

It is good to remember that the process of choosing and working with crystals is a mutual decision; you not only choose the crystal—the crystal chooses you! The very fact that you are even reading this book is because, consciously or unconsciously, you have heard the song of the crystals calling your name.

A great many of the people who have come to me over the years seeking crystals have come out of almost a compulsion to become involved in their use—even though consciously they know very little, if anything, about them. Quite a few have recounted amazing and magical dreams, in which they saw the very crystal which they then found weeks, months, even years later!

For many years, while in a meditative state, I would have the same vivid, recurring vision, which would leave me feeling very strong and cleansed and full of joy. In the vision, I would be walking alone, under a bright turquois sky in an almost blinding light, among rolling sand dunes. These golden dunes stretched for miles, as far as I could see. The air was very still, and I could feel the warm caress of the sand beneath my feet. Suddenly, an incredible unearthly music filled my being. Out of the sand all around me, rose thousands of perfectly exquisite crys-

tals, all sizes and shapes. They radiated a surreal light of many different hues—violets, purples, ruby reds, emerald greens, electric blues—constantly changing and pulsating, dancing all around me. I felt an overwhelming Love. An awe and ecstacy filled my spirit, which remained with me for weeks.

I have been drawn to the beauty and the magical qualities of the mineral kingdom since I was a child, and on many, many occasions, during deep meditation or in vivid dreams while sleeping, have been given great jewels of Divine Truth by beautiful Light Beings. (These Beings are constantly there for every living thing, providing guidance to anyone who is ready to receive it.)

Many of these truths have to do with the connectedness of man to all other life forms in the Universe—of which the mineral kingdom is a part—and the ways in which man may utilize this connectedness to raise, purify, unify, and harmonize the vibrations of all creation to enable us to grow into our Divinity.

As we work with the crystals to heal ourselves and others through Divine Love, we are also healing the planet—Home, Sweet Home for now. It is past due. The Earth has given and given, lovingly and unselfishly. She has been greatly abused by man in his ignorance—her gifts squandered and grossly misused. Time now to give something back. Time to give Love, Understanding, and Respect back to the Earth. TIME TO GET CONSCIOUS!!!!

I feel very blessed to be given this gift to give to you. It is my hope that you will take this message of Love into your hearts and then pass it on.

3

CHOOSING A CRYSTAL

Some of the things to consider when choosing a crystal are: what you wish to use it for, clarity, shape, size, and—last but not least—the feeling you get from it.

Indeed, feelings—or intuition—are probably the best guide in the process of selection when it comes to crystals. Although visually you may be drawn to the most aesthetically formed crystals, it is often the less obvious, more subtle ones that you end up taking home.

CLARITY

The clearer the crystal, the more rapid and unimpeded the energy flow. In this respect, they are like man. So generally, one of the first things you want to look for is clarity.

However, there are times when you will want to use crystals containing some milkiness, or cloudiness. This cloudiness occurs naturally in most crystals, usually at the base.

Cloudiness will give a crystal a more gentle, grounding quality, and is especially useful when dealing with spaciness, in yourself or others. It helps to ground the Divine energy channeled by the crystal in this physical plane.

For instance, if you were working on healing someone who was severely unbalanced, mentally or emotionally, some cloudiness would be necessary. Using a perfectly clear crystal at this time could prove to be dangerous, possibly having an extreme shattering effect on the patient.

On the other hand, when working to raise vibrations from a more to a less dense level, or when attempting to remove blocks in certain centers of the body, you would want as much clarity as possible. The perfectly clear crystals are, for this reason, excellent meditation enhancers.

When you want to achieve a perfectly balanced state—comfortable, grounded, and aware in **this** reality—a crystal with a balance of cloudiness and clarity is an excellent choice.

INFRACTIONS

Many crystals contain infractions, some of which cause incredible rainbows to form within their depths. It is the crystals containing these infractions which are excellent dream and vision enhancers.

We spend so much of our time while on this planet unconscious, and therefore unable to learn the lessons we came here to learn. Crystals can help us to function in a more aware way, not only in our everyday lives, but also while in the dream state.

When placed by your bed at night—or, better yet, under your pillow—they will not only increase the vividness of your dreams, they will also enable you to remember them more clearly. You will become conscious during the dream state, controlling your reactions and actions while dreaming.

Gazing into crystals containing rainbows will greatly increase your ability to envision, as well as get you in touch with your intuitive self.

There are also crystals which contain thin straight criss-crossing lines of other minerals. These are called rutillated quartz crystals, and are particularly powerful. The inclusions will often increase the conductive quality of the crystal, as well as its ability to hold a charge.

SIZE

The size of the crystal used, strangely enough, does not seem to have an effect on the strength of its vibrations, or its ability to heal. Some of the most effective crystals I have worked with have been very small.

However, as a general rule, the ones that are best used for healing are usually three to ten inches long, and one to three inches in diameter.

If they are to be used as a generator of energy on a large scale, or to protect an area, the larger ones will do best. The size of the crystal ***does*** *seem to have some rela-*

tion to the size of the area influenced. Simply placing them in a room, or at the four corners of a certain area, will create a protective spiritual atmosphere conducive to awareness, and clarity of purpose.

The smaller crystals are excellent to use as healers, generators of energy, and protectors when worn as pendants or in a pouch around your neck, or simply carried in your pocket.

The best crystals to use for meditation would be either a size that will fit comfortably in your hand, or you may prefer to use the larger generator crystals to create an area conducive to this purpose.

SHAPE

Crystals come in every shape and form imaginable—some are almost ***unimaginable!*** *Each one is unique, with its own personality, its own gift to offer to the world. Some have chosen to be specialists, focusing on one particular aspect of the Divine.*

Quartz crystals are, with few exceptions, all six-sided, with a cap-like point on one or both ends of the shaft.

It is best to use crystals with at least one intact point and no ***severe*** *chips. Small chips, and indentations where the crystal may have accomodated itself to, or grown* *around, other crystals during its formation will not inhibit its power in the least.*

Crystals with a completed point at each end are known as double-terminated crystals. These are especially powerful, as they contain within them a completed electro-magnetic circuit. When you are holding a crystal with only one completed point, then* **you** *complete the circuit.*

This is good to remember when choosing a healing crystal. Double terminations provide more protection for the healer because of this completed energy circuit. You direct and add to the energy to be directed—as when using a single terminated crystal—but the energy will not pass through you unless you so direct it—and then only if you feel strong enough to handle any imbalances which may flow from the patient to yourself.

Twinned crystals—crystals which share a side, or that have two or more points on each end—seem to be able to channel twice as much energy. They also have other qualities, which are too subtle to translate into words, but which will definitely be felt by the user.

Crystals with points containing six faces of approxi-

**Crystals with an incomplete point—caused by breaking them off of the matrix in which they have grown—will eventually heal themselves by forming a completed point at the broken end. Occasionally, two crystals will grow together while in matrix, also forming a double termination.*

mately equal size and shape are very balanced. They channel the many aspects of Divine energy equally, not concentrating on any one particular quality.

Crystals with broad, shovel-shaped points containing one large face, the other five very small—if not infinitessimal—are known as third-eye crystals. They seem to be particularly effective in awakening and strengthening the varying psychic abilities latent in man: clairvoyance, intuiting, healing on an astral level, astral projection, etc. They enhance the ability to envision the Divine, and then to project that Vision into this physical reality, healing and balancing the imbalances—creating Divinely.

This gift should ***never*** *be used for any purpose that is not Divinely inspired. The crystals will reflect back onto you any imperfections you send out—they* ***will not*** *accept these! You will harm only yourself, sometimes causing great pain and devastation.*

INTUITION

Your higher self—whether you are consciously aware of it or not—always knows the truth of what is perfect for you. For this reason, when it comes right down to it, feelings are the best way to zone in on the crystal that is right for you.

The best way to tune in to these feelings is to hold each crystal in your left hand (your receptive side), and then sit quietly, eyes closed, attempting to merge your energy with the energy of the crystal. You may feel a slight electrical charge—a tingling—in your hand, or running up your arm. Or you may feel a pulsating energy. Sometimes there will be a sensation of heat or expansion in some part of your body—particularly in the area of your heart, your solar plexus, up and down your spinal column, or in the center of your forehead.

Your body will let you know, more often than not, when you have picked the right one.

If you do not experience any of these subtle—and sometimes not so subtle!— physical signals, you may want to look at other things.....such as the fact that, even though you think you want that large crystal with perfect clarity, you keep picking up the one with a little cloud at its center!

Trust your heart, not your head, and you will never go wrong!

4

THE DEDICATION

After you have chosen your crystal, and before you begin to use it, you must perform a dedication, consecrating it so that it may only be used in the service of God, instead of being manipulated solely by the unbridled will of man—untempered by Divine Will, Wisdom, and Love.

Take your crystal to a place where you feel comfortable and safe and the most at one. Your regular meditation spot, or an outside location, among the mystery of nature, are excellent choices. Playing beautiful music at this time will add greatly to the meditative atmosphere.

If you feel more comfortable with ritual, you may want to form a circle of seven white candles around you, placing one more before you in the center of the circle. The light of these candles symbolizes the Divine Light that surrounds and permeates the Universe. When this Divine energy is allowed to flow unimpeded, no imbalance or inappropriateness of any kind can exist.

Now sit or stand, holding your crystal before you in both hands, and after moving your consciousness into a meditative state of expanded awareness, visualize a protective sphere filled with bright White Light surrounding you, top to bottom and all around. Then, with your mind's eye, see this same type of sphere around your crystal.

You now have your crystal surrounded and protected by its own sphere of impenetrable White Light, within the larger sphere which surrounds and protects you. Your energies are merged with the energies of the crystal. You are one at this point. Nothing of an inappropriate nature can exist within this sphere of Divine Light.

Now declare, aloud if possible:

"Only DIVINE LIFE, DIVINE LIGHT, DIVINE LOVE, DIVINE JOY, and DIVINE WISDOM can be here. Only these can flow from, or come to me and this crystal. All inappropriateness and imbalance will be healed and made perfect through the energy of DIVINE LOVE channeled through the crystal I hold in my hand. This crystal and I are now clear, pure conduits for the Divine. I shall see with the eye of God, hear with the ear of God, know with the mind of God, speak with the voice of God, and love with the heart of God. Thank You, GREAT UNIVERSAL HEALER, for making it so."

Every time you use your crystal—for that matter, every time you meditate—you should repeat this dedication to the Divine, in whatever form feels most comfortable to you. In this way, you clarify the fact that the only thing with any validity and reality for you, the only thing recognized as appropriate, is DIVINE ENERGY. All else is illusion, and should not be given even the pretence of being real.

When not using your crystal, keep it in a pouch, or wrapped in a cloth of natural material, such as silk, cotton, or velvet. If you have more than one crystal, keep them wrapped individually.

The more time you spend with your crystal, the better. Carry it with you during the day if possible, and sleep with it at night. It will rapidly become a part of you, opening up many doors to incredible powers and wondrous realities that are your Divine birthright.

5

CLEARING TECHNIQUES

Quartz crystals protect and heal not only by generating and transmitting intense amounts of pure Divine energy, they will also absorb into themselves any impurities within their vibrational range, and then transmute that energy. They have the power to clear themselves, just as man has the power of self-healing.

However, like man, if there is constant stress (subjection to overwhelming imbalances), with no time to recharge, the healing takes place slowly, or sometimes not at all. The imbalance is allowed to build up.

For this reason, it is good to clear your crystal periodically, especially when it is being used in the healing of others.

There are several techniques for doing this. They are all very effective. The best way to choose which to use is with your intuition. I have tried all of them at one time or another, and personally prefer clearing by channeling Light; the sun; burying the crystal in the ground; and immersing it in sea-salt and water, in that order.

CHANNELING LIGHT

Sit in a quiet place, preferably one that has been consecrated for meditation or other spiritual pursuits. Crys-

tals respond to music, so it would help to have some astral-type melodies playing during this process (it wouldn't hurt you, either !).

After you have done the preliminary work clearing and balancing yourself (see section on meditation techniques) then you are ready to become a conduit for channeling pure Divine energy.

Pick up your crystal, holding it in front of you, a hand on each end. Begin a build-up of loving Divine energy in your heart center. When this connection has been fully established, and the energy can clearly be felt, pulsating strongly, direct this Divine Light from your heart center through your right arm to your right hand, then through the crystal to your left hand, down your left arm, and back to your heart, so that a complete circuit is established. Visualizing the energy as a flow of brilliant golden-white Light sometimes is helpful in this process.

Continue channeling for at least twenty minutes. Then place the crystal in your right hand and form a protective sphere visually around yourself and the crystal, declaring, "Only Divine Light, Love, and Life can be here; all else is transmuted into Divine Perfection."

SUNLIGHT

Crystals love the sun, just as all other forms of life

do. It has a very cleansing and generating effect on them. Indeed, this particular method of cleansing is also one of the most excellent ways to build and strengthen an energy field in your crystal.

First, bathe your crystal in distilled or spring water. Then, leaving it wet, place it in **direct** *sunlight for three days (this does not mean through a window, as the full range of the sun's rays will not penetrate glass.) A spot with a southern exposure, out of the shade, is best. Leave it out at night also during this period; lunar energy is another good balancer.*

It is beneficial to use the same spot for this process whenever possible, consecrating it for this purpose.

A variation on this method is to use colored glass containers, putting your crystal in them, and then placing them in the sun for seven days. The crystal will become attuned to the particular vibrational qualities inherent to that color (see section on chakras).

However, this method is more for charging...it is not as effective as a clearing device.

RETURNING CRYSTALS TO THE EARTH

The Earth is truly home for crystals; the place where they were born, a sanctuary where they can recharge in stillness, within the comforting arms of their mother.

The best way to carry out this method is to first find, consecrate, and then carefully prepare a place within the earth in which the crystal will be at home, and can be left to recharge and realign itself with the natural order of things, which the Divine, with great Wisdom and Love, has provided for planet Earth and all who reside here—in whatever form we choose to manifest at the moment!

After you have found a spot which feels right, and consecrated that spot, then dig a hole, preferably rounded, at least $^{1}/_{2}$ to one foot deep. Then line the sides and bottom with rock (sandstone, quartzite, or marble would be best), or you may use copper plating, if you wish. Then find a large stone to cover the top of the hole. Pour distilled or spring water over the crystal until it is thoroughly wet, and place it inside this space, cover with the top stone, and leave it, undisturbed, for three days.

To make this process even more effective, you might want to place an **open** pyramid—which can easily be constructed from copper tubing—directly above your crystal's sanctuary.

After the three days are up, go to the spot, remove the crystal, again bathing it in clear water, thank the Divine, and welcome your friend back.

SEA WATER

One of the most healing and rejuvenating experiences for man is to bathe in the clear, refreshing water of the sea. So it is for crystals.

Part of this rejuvenating quality comes from the salt content of the ocean. Salt is crystalline in nature, and it will pull negativity and toxins out of the bather and into itself.

Obtain a container large enough to hold at least one quart of water. A fish bowl is an excellent choice. If you live near the ocean, you are in luck, because you have an endless supply of sea water; if not, you can make your own, using pure sea salt (which can usually be obtained at your local health food store), and distilled water.

Mix the water with two teaspoons of sea salt per quart. Then place your crystal into the container of water. Do not cap the container. If you so choose, you can place it in the direct sun to further enhance the effectiveness of this method.

Leave your crystal in the solution for from one to three days. Then remove it and rinse it with clear water. Do not save the solution; with each cleansing, it is best to use fresh sea water.

CLEAR RUNNING WATER

For this method, you will need to have a natural source of pure, running water—such as a spring or a mountain stream—nearby.

Make a container for your crystal out of open weave netting, or you can make a type of cage from copper wire.

Find a spot that feels right, and then place a post deep into the bank of the spring or stream. Next, tie a sturdy piece of string or strong fishing line to the net or cage, and fasten it to the post.

Place your crystal in the container and then gently lower it into the water, making sure there are no rocks which it might hit against and, in so doing, possibly sustain some damage. Then leave it for three days undisturbed, after which you can remove your crystal, thanking the water for it's help in the cleansing.

MAGNETS

When crystals become unbalanced, one of the things that happens is, their electromagnetic field gets confused and somewhat deformed, and their molecular structure gets knocked out of alignment. Magnets will help to realign this energy field.

Take two magnets, Place your crystal on a cloth of a natural material, such as silk or cotton, in a consecrated spot. Then arrange the two magnets at each end: positive at one end, negative at the other. Leave the crystal and the magnets in place for three days so that they can accomplish their task. Use the same magnets each time, and when not in use, keep them wrapped in the cloth.

6

ESOTERIC ANATOMY

Although crystals are of themselves powerful generators of energy, as transmitters and directors of that energy, they will only be as effective and powerful as the person using them. That means you.....YES, YOU!!! It's time to get to work clearing out your own blocks.

However, a most wonderful thing happens when you are working with crystals: the more you work with them, the clearer you become; the clearer you become, the stronger the flow of energy and the more effective the crystal; the more effective the crystal, the stronger and more direct your energy, and so on.....in an unending spiral toward becoming a pure channel for Divine Light on this planet, throughout the Universe, and beyond.

There are many meditations that can help you to get in touch with your Divinity. I will give you a few in this book. But first, since many of these techniques require visualization and a rudimentary knowledge of the different energy centers of the body, the colors associated with these centers, and the different bodies which make up your being, you are now going to be presented with a crash course in esoteric anatomy.

CHAKRAS

Chakras, or energy centers, exist throughout your body. There are seven main ones, corresponding generally to the seven major groupings of endocrine glands and neural concentrations.

Each of these chakras has a particular vibration, producing its own color which can be seen in the aura of man (the energy field generated by your being in the physical plane).

These centers are aligned roughly from the base of the spine to the crown of the head. And the colors are incredible, magical.....the colors found in the rainbow, in all their varying and subtle translucent glowing hues.

At the base of the spine lies the root chakra, in which man's sexual energy resides. This is creative energy in its most physical manifestation. The color corresponding to this center is a deep, rich, clear red.

Just above this, directly below the naval, lies the chakra which corresponds to the digestive and eliminative centers in man. It is through this chakra that we process life—accepting or rejecting. The color of this particular vibration is clear bright orange.

Between the naval and the diaphragm lies the solar plexus center. It is through this center that man expresses strength or weakness of will. The etheric body is attached to your being at this point; therefore, it is usually here that trauma or shock on other levels will first manifest themselves physically. Intuitive feelings are often felt here first. The color of this vibration is a warm yellow.

In the center of your chest lies your heart chakra. This is the seat of your emotional self—your Love center. The color is a clear emerald green. Great healing—of yourself and others—can take place through this center—the area in which the upper and lower bodies come together.

At the throat lies the creative center, where man expresses all that he is verbally; the area at which ideas can be manifested in physical form. The color corresponding to this particular chakra's energy is sky-blue or turquoise.

In the center of the forehead, just behind and above the eyes, lies the third-eye chakra.....seat of the soul in man; the place at which the astral body makes itself known. The wisdom of the Universe is available to you through this vibration. The color here—at the Eye of God—is an electric indigo blue.

At the top of your head lies the crown chakra, the place where man and his spirit are one. The vibration creates a clear, strong violet color.

Above your head, out of your physical body, yet still within your energy field, lies the chakra which connects you to your Christ-self…the place where the Divine Creator within you first manifests itself. The color is a brilliant, magnificent golden glow.

Although these colors originate in their respective centers, when you are clear and balanced, they permeate all parts of your energy field, creating an ever-changing, incredibly exquisite kaleidoscope of hues pulsating around and through you.

As you become clearer, you will be able to see these colors emanating from everything. It's easiest to see them with slightly unfocused eyes, at the periphery of your vision.

When there is any weakness or cloudiness, and sometimes even an absence, of a certain color in a person's aura, it will often indicate a block or imbalance in that part of their being. There will usually be one or two predominating colors in an individual's aura. These colors are not necessarily constant; however, they do correspond to the areas of your being through which you are manifesting at the time.

All of life is full of just such mystery and beauty. This is only a small part of that mystery, available to all who are brave enough, and free enough, to allow themselves to accept the way of the Divine Mystic within.

THE SEVEN BODIES OF MAN

Man has seven basic bodies through which he manifests energy.

The first is the physical body. It is the body we are most familiar with, the place at which the vibratory rate is the densest, the one through which we manifest on the physical plane.

The second is the etheric body. This is the energy body directly surrounding and permeating every cell in our being. It is the place where disease or fear or any other weakness or imbalance will first manifest itself. The strength of the physical body is in direct proportion to the strength of this body.

The third is the emotional body: the place where man manifests himself in an emotional way. Love resides here, and a great capacity to heal on all levels.

The fourth is the mental body: the place where man ideates, and then creates.

The fifth is the soul body, or astral body, of man. All things learned from past-life experiences, all karmic ties, all that has happened, or that will happen to this being on an ***individual*** *level is stored here.*

The sixth is the spiritual body. This is where Divine wisdom on a Universal level is made known to man. Here individuation falls away, and a oneness with all creation is experienced.

The seventh is the Christed body of man: the place at which he becomes God the Creator. It is from this center that Divine Will is able to manifest—to create perfection where there was imperfection, health where there was disease, balance where there was imbalance, and Divine Love which turns all wrongs into rights.

When you can remove all blocks in every body in your being, then this Divine energy will be free to flow throughout all parts of you, uniting and dissolving all boundaries. You will be able to accept, ***without question,*** *the Truth of your Divinity.*

7

MEDITATIONS

Place your crystal in your left hand, point toward you, to receive energy; in your right hand, point away from you to send energy or to charge your crystal with a particular vibration.

Before you begin any meditation, it is important that you do a brief protection. This will ensure that only Divine Love and Light will be working through you.

RAINBOW BRIDGE MEDITATION

This meditation will intensify the vividness of your visions, increase your ability to perceive auras, as well as enable you to locate any blocks in your being and then dissolve them.

You are going to visualize the color of each chakra, starting with red, then orange, yellow, green, blue, indigo, violet, then gold, into white.

As you visualize each color, see it first swirling all around you, throughout your being. Then ***become*** *the color. Feel it in every cell of your body. Stay with it as long as it feels right, and then see it spiraling down to rest in the appropriate energy center. Feel it pulsating there, growing stronger. Then, when this vibration is firmly rooted, move on to the next color, and so on, up the rain-*

bow ladder, until you reach the golden light.

At this point, your consciousness lies outside of your physical body. You are ready to slip into the Divine White Light—the place where there is everything, and nothing. Stay in this expanded state of awareness, empty and open, receptive to the Divine Wisdom and Love which is here for you.

Then, when you feel it is time to return, go slowly back down the ladder—gold, violet, indigo, blue, green, yellow, orange, to red—again seeing and feeling each color in the appropriate area of your being.

Now let the Divine White Light flow into your body, filling your whole being. Let it, and with it, your awareness, radiate out from you in an ever-expanding circle. You are a part of all that is, and all that is, is a part of you. This meditation will help you to re-establish this connection, and to carry it with you always.

Remember during this process to breathe deeply. Breath is life.

If you notice any color that is hard for you to visualize, or any place at which it is hard for you to feel this energy flow, that is more than likely a place where a block exists. Make a point to work on that color in particular-.....wear it, surround yourself with it, think it, whenever possible. You will find the block will soon disappear!

LIGHT MEDITATION

During this meditation, it is best to hold your crystal, point in, in your left hand.

The purpose of this meditation is to increase the clarity and unity of man's many selves.

First, do a protection.

Then, center yourself in your physical body. Really feel the energy there, visualizing it as a brilliant golden light, filling every cell. Then state: "Only Divine Clarity and Perfection can exist in my physical being."

Now move on to your etheric body. Really feel the energy there, and experience this part of you. Center the Golden Light in this body; watch and accept anything that it illuminates. Again state, "Only Divine Clarity and Perfection can exist in my etheric body."

Do this for your emotional, mental, soul, spiritual, and Christed bodies, becoming totally conscious and aware of these aspects of your being.....what they feel like, what the light illuminates for you there. If you see imbalances, fears, or other impurities in your mind's eye, then see the Light melt them away.

It is important at this point to note that any time you are faced with an imperfection in yourself, it is your responsibility as a Christed Being to first de-fuse, or balance, that imbalance before you release it. If you do not deal with it yourself, you are simply dumping garbage into the astral plane, cluttering up the airways with imperfect thought forms, which have the potential to harm you or someone else. Thoughts are energy—it is up to you, before you release your thoughts, to make sure you do so with the statement that they are being balanced through Divine Gentleness and Love.

Now see and feel this golden Divine Light flowing freely, strengthening and cleansing, uniting every aspect of your being. Feel how Infinite you are...how empty, yet how full. This is Who You really Are!!!

CHARGING YOUR CRYSTAL

For this meditation, you will want to hold your crystal in your right hand, point out.

Do your protection, and then allow yourself to dissolve into the Golden Light. Become that Light, pulsating, glowing. You are now an expanded being, containing within yourself the Universe. Feel the power and glory, the wonder and mystery, of what you ***truly*** *are!*

Then, pull this pure Divine energy together at your heart center. With every pulsation, feel it flowing down your right arm into the crystal.

If you specifically want to fill your crystal with healing energy, see this golden light, like a brilliant sun, ema-

nating from your crystal.

If you want to use your crystal to enhance creativity, fill it with bright turquoise.

If you wish the purpose of your crystal to be to enhance the clarity of your third-eye vision, see it filled with electric indigo.

If it is Divine Love—a deep ruby red, or soft pink. Feel the Love...feel it flowing.

If you want to establish a good balance of all of these aspects of the Divine, fill it with the rainbow colors. As you do this, you are also charging yourself with these particular aspects!

Remember, at the end of each meditation, thank the Divine Energy within yourself for being.

CRYSTAL MEDITATION

Hold your crystal in your right hand, point out.

Do a protection, and then the rainbow meditation, until you reach the Divine White Light. In this expanded state of awareness, open your eyes and really look at your crystal. Note every facet, each infraction, all rainbows or clouds within it.

Now close your eyes and feel yourself as energy flowing into the crystal, until your consciousness is inside of it. Look around. See your crystal from the inside out. Ex-

pand your consciousness until you are no longer you, but your crystal instead.....a perfect faceted jewel containing rainbows and clouds, reflecting light, pulsating with energy and joy!

Feel as your crystal feels; experience reality as a crystal. Then recognize that this crystal reality is the same as your reality.....you ***are*** *a crystal, in another form. You, in your humanness, also reflect light, pulsate with energy, and contain incredible light shows.*

It is the same energy, in different forms! Yet the forms are just that...forms created by the Divine to express the many aspects of Divinity. You and your crystal are truly one.

To close, come into your physical awareness by going back down the rainbow ladder.

8

HEALING TECHNIQUES

Before beginning these, or any other healing processes, it is wise to first do a clearing meditation, so you will be as clear a channel as possible. The rainbow meditation or the light meditation included in this book are both excellent methods to achieve this clarity.

Once again, it is also important to do a protection at the beginning. This is especially true if you are healing another.

HEALING YOURSELF

Place your crystal in your right hand, point out.

After doing a protection and a balancing meditation, visualize the golden Christ Light surrounding and permeating you. Then scan your entire being, becoming aware of any places in yourself where the Light is having a hard time getting to.

These places will usually, but not always, correspond to the place where the disease or imbalance is located. (Sometimes, even though we are experiencing pain in one place, the cause of this discomfort is rooted in another place, or part of our beings, such as the emotional body, or the soul body.)

Visualize and feel this Golden Light building in your heart center. Then, with each pulsation, send it down your right arm and into the crystal. There it is further amplified. Now direct it back out the crystal point to the part of your body where the block is located. Or, you may simply let the energy flow into the crystal, circulate there, and then flow back down into your heart center and from there on to the area where it is needed.

See yourself as perfect, feel your perfection, and know that anything else is illusion caused by a lack of acceptance of your Divinity. It has no place here. It does not exist.

HEALING OTHERS

Sit facing the person to be healed. If this person is not with you physically, then visualize him in your mind's eye, and after asking his permission to proceed, then you are free to continue. Do a protection for the both of you, stating it out loud.

*With your crystal in your right hand (point out), visualize Divine Golden Light around both of you, permeating your beings. The places of imbalance will again usually show up as dark places, or places where the light does not seem as strong.**

**It is important to emphasize here that there are many ways to discover the site and cause of imbalance other than visualizing. After merging with your patient, you may feel within your body phantom pains...echoes of the discomfort felt by the one to be healed. Or, you may feel any imbalanced areas as inordinately cold or hot places within the energy field of the patient, depending on the problem. As you work with this energy, you will come to "just know" what the problem is and how to fix it...not all of the time, but a great deal of it.*

Using your crystal, direct this golden light of Divine Love and Perfection into these places of imbalance, balancing them and restoring them to wholeness.

Do not worry about being depleted. There is an infinite supply of this Divine energy. When you are in an expanded state of awareness, you have removed all the blocks that would prevent you from experiencing the full flow of this energy.

See your brother as whole; do not accept him as anything less.

Another variation on this method is to stand behind your patient, and direct the energy to the different parts of his being by placing the crystal directly on the body or the in surrounding aura where the healing energy is needed.

THE CRYSTALLINE STAR

This method requires six crystals, and is best done out of doors.

Find a spot that feels right and consecrate it.

Then, holding the six crystals—three in each hand—do a protective meditation, and dedicate the following act to Divine Love and Healing.

Place the crystals on the ground so that they form a star out of the joining of two triangles.

Three crystals—the three forming the triangle whose apex you are facing—should sit with their points toward the sky. The other three—forming a triangle with it's apex directly behind you—should be placed point down. Now you should have the shape of a star with yourself in the middle.

Sit or stand, which ever is more comfortable, and then do a healing meditation for yourself; or, if you are healing another, for that person—who should also be within the star. If they are not present, visualize them in your mind's eye. Always seek your patient's permission to proceed before continuing with the healing.

Keep the crystals used in this meditation together, separate from your other crystals, and clear them thoroughly after every healing.

9

COMBINING OTHER GEM ENERGIES WITH QUARTZ CRYSTALS

When other gemstones are combined with quartz crystals—either mounted together and worn as a pendant, or simply held in the same hand or placed together in a pouch—then the specific qualities of that gem are taken on, magnified, and transmitted by the crystal...that marvelous cosmic amplifier.

Every member of the mineral Kingdom is crystalline in nature, belonging to one of seven major crystalline structure groups. As crystals, they each have their own specific characteristics...qualities that have an effect on the world around them, on all levels of existence. Some, however, are more effective in relation to man at this particular stage of his development than others.

The color and clarity of the gems is a definite factor in their degree of effectiveness.

Every color in the spectrum of colors, visible and invisible, has its own vibration, which corresponds to the vibration of certain qualities and aspects in man's being (see section on chakras, and the seven bodies of man).

The clearer and deeper the visual color emanating from a particular stone, the stronger and more defined the corresponding vibration.

I have found the following gemstones to be among the most beneficial, their vibrations the most conducive to the growth of man into his full potential as a whole, vital, Christed Light being. There are many more in existence on this jewel box of a planet...some still waiting within the earth's velvet folds to reveal themselves at the proper time.

AGATE

The colors found in agate range from browns and greens to blues, pinks and purples. It is an opaque to translucent mineral, effective when cut into squared or rounded shapes, and mounted in silver.

This stone is an excellent grounder, providing a solid physical base through which the more subtle vibrations of the body can flow. It is a stabilizer of the emotions, clarifying the truth in a situation, and then making the acceptance of that truth easier for the wearer.

Fire agate—a deep red with a firey sheen—serves as an energizer. Moss agate—a variety of this mineral containing moss-like green and brown fragments—forms a strong connection with the nature spirits, or the devic kingdom. It is especially beneficial to those who work with the plant kingdom.

Agate combined with clear quartz gives the wearer a stable footing and connectedness with the Earth, and the calmness and assuredness to deal with any problem that might arise.

ALEXANDRITE

Alexandrite is a very rare clear gem containing changing colors of red-violet and emerald green, depending on the light. It is best cut into rounded shapes (although square is sometimes good), mounted in white or yellow gold.

This stone has an alchemical, ceremonial, highly mystical quality, and it will transmit these qualities to the wearer. It transmutes energies on an emotional, physical, and astral level, healing any imbalances that may exist at these levels. It especially aids in the combining of the spiritual, emotional, and physical aspects of love, bringing an experience of wholeness at the heart center. It is also a transmitter of energies at a very high rate of speed.

When worn in combination with quartz crystal, the wearer becomes a great, clear channel for healing through Divine Love on all levels.

AMBER

Amber is a stone of organic origin, the fossilized resin of prehistoric conifer trees. This translucent to opaque golden gem can be cut into any shape, mounted in silver or gold.

Amber has a calming, soothing effect on the emotional and physical body of the wearer. It is highly absorbtive, especially at the etheric level, removing any inappropriateness trying to enter the body at this point; thus protecting and strengthening the health aura. Because of it's absorbtive quality, it should be cleared often.

When used with quartz crystal, it protects the physical and emotional health of the wearer, imparting great strength and serenity.

AMETHYST

Amethyst is a form of quartz crystal, and is found in all shades of lavendar, violets, and purple. The deeper purples are most effective, faceted and cut into squared or rounded shapes, mounted in gold, although silver is also acceptable.

Amethyst is a most powerful transmuter and purifier of energy from a lower to a higher, spiritual level. It will absorb and rebalance imbalanced energy, thus serving to protect the wearer from any inappropriateness, as well as helping to deal with these imbalances within the self.

This stone generates its own energy, cutting through denser matter to stimulate and strengthen the spiritual nature of all things. It is a great healer.

When worn in combination with clear quartz crystal, the amethyst helps the wearer to be constantly aware of the spiritual aspects within himself, and the spiritual implications of his various experiences while on this planet. It also will strengthen any healing abilities inherent in the wearer, as it aids in transmuting all energies from a lower to a higher, purer vibrational intensity.

AQUAMARINE

Aquamarines are blue, green, or clear transparent gems, best when faceted, cut into any shape, and mounted in white gold. The bluer stones are best.

This stone has a direct effect on the emotional and mental centers of man. It balances and purifies, and smooths the way for the direct manifestation of spiritual energies in a creative way on the physical plane. It also enhances the meditative state in the wearer, centering awareness in the spiritual and higher mental planes of existence.

When worn with clear quartz crystal, aquamarine nourishes the spiritual, creative, aspects of man, clearing the way for a direct understanding and manifestation of the Divine in action on the physical plane.

AZURITE

Azurite is an opaque, copper-based mineral found in shades of electric blue to indigo. It is most effective when cut into flat, squared shapes, mounted in silver.

This stone works on the physical and emotional bodies. It brings calmness and clarity of awareness to the wearer. It greatly enhances the meditative states, stimulating and awaking the third eye energies. Any impurities on a physical or emotional level will be drawn out of the body of one who possesses this stone.

Azurite, when combined with clear quartz, will draw out toxins and impart clarity and stability to the wearer.

BLOODSTONE

Bloodstone is an opaque, deep green mineral containing bright red specks. It is most effective when cut into a domed oval shape, mounted in silver.

This stone is one of the most powerful movers of kundalini energy when used by a clear channel. It is also an extremely strong healer on a physical level, especially in cases of internal bleeding. Because it is such a powerful stone, it should be used with extreme caution and complete awareness.

When used with clear quartz crystal, it transforms the wearer into a conduit of healing energy which can then be used on oneself or others to remove any blocks pre-

venting the smooth flow of the kundalini energy, as well as repairing any imperfections or illness on a physical level.

CARNELIAN

Carnelians can be red, orange, or yellow-brown. They are opaque or translucent minerals, and are most effective when of a deep red-orange color, cut in an oval or square shape, mounted in silver or gold.

This gem creates it's own electro-magnetic field and is one of the most potent energizers known to man, affecting the kundalini, causing it to rise, shaking loose any blocks along the way. It works mainly on a physical and astral level, serving as a filter to weed out any toxins and imperfections. It connects the wearer's consciousness to his soul body, making past-life experiences and present purposes clear.

In combination with clear quartz, carnelian serves as a stimulant, an energizer, and a shatterer of any blocks or misconceptions in the wearer. It is VERY POWERFUL and should be used with care and awareness.

CHRYSOPHRASE

Chrysophrase is a translucent to opaque apple green form of onyx, best cut into a flat square, or rounded dome shapes, mounted in silver.

This gem is an excellent balancer of energies. It will pull back into alignment anything out of line, or otherwise inappropriate, on a physical, emotional, or mental level. It also brings these centers into direct alignment with one another so that the Divine energy may flow more freely between them. It has a strong healing effect on the heart center, bringing peace to the wearer.

When used with clear quartz crystal, it strengthens the **auric** *field, heals any imbalances present on all levels, (especially the physical, emotional, and mental), and promotes a strong sense of well-being in the wearer.*

CITRINE

Citrine is a type of clear to translucent quartz crystal that is brown, golden, or light yellow in color. It is most effective when faceted, cut into any shape, and mounted in silver or gold.

Citrine is a grounding gem, working mainly on the physical and emotional bodies. It strengthens the will and the solar plexus energies, balancing these areas. It is an expansive stone, opening the wearer to new ideas, broadening one's view of reality.

When combined with clear quartz, citrine imparts the qualities of emotional warmth, openness, and physical strength.

DIAMOND

Diamonds can be clear, yellow, blue, pink, or green. They are the hardest mineral known on earth, metamorphosed by tremendous pressure and heat. They are a symbol of man's ability to pass through the fire of learning to become a perfect, glistening jewel in the crown of God.

Diamonds are strongest when faceted and cut into a rounded shape, though other forms are also effective. Golds of any type are the best mounting material.

These stones are one of the strongest magnifiers and transmitters of energy, capable of creating an invulnerable protective field around the wearer. However, they are also one of the most absorbtive gems, indiscriminate as to what type of energy they take in and give out. This means that they can be tools for great good or tremendous evil, depending on the wearer. They will take on and magnify the qualities of the owner, including any trauma that person experiences, and magnify them. Once imprinted with a certain quality, they are extremely difficult to clear.

If diamonds are used in combination with quartz, it is best to use amethyst or rose quartz, instead of the clear variety. That way, the diamond will take on the qualities of spiritual transmutation and Universal Love engendered by these stones.

EMERALD

Emeralds are a deep, rich, vibrant green. The stones selected should be as clear as possible, preferrably faceted. They are most effective when cut in a square or rectangular shape, mounted in yellow gold.

The heart is especially affected by this gem. It has the quality of balancing and healing the emotions on a physical level.

Through its loving vibrations, it gently combines the lower creative sexual energies with man's higher mental creativity so that it can be incorporated as wisdom and expressed as love in the total sense of the word.

If there are any blocks or imbalances, especially in the area of the heart, it helps to remove these, and then heal and re-balance.

Wearing the emerald in combination with quartz crystal will encourage the qualities of wisdom, givingness, open lovingness, understanding, and strength.

GARNET

Garnets are clear, transparent, or translucent stones found in reds, violet-reds, and greens. They are most effective when faceted and cut into rounded shapes (although square shapes are sometimes good), and mounted in yellow gold.

Green garnets work on a more physical level, while the red tints have a greater effect on the emotional and spiritual bodies. They are both great healers of the heart, and stimulators and healers of third-eye energy. They are clear conduits for the passage of energy to or from the wearer, with the added benefit of first removing any inappropriate thought forms. They enhance love on a Universal, devotional level, bringing clarity and a sense of purpose to those who possess this stone. They aid in the grounding of spiritual, psychic energy in the physical body.

When combined with clear quartz, the garnet becomes a catalyst for great love and peace and awareness, as well as imparting an ability to move surely and freely through the astral realms.

JADE

Jade is found in many colors, among them green, pink, lavender, white, yellow, and red. It is an opaque stone, best cut in any shape appealing to the wearer, mounted in gold or silver.

Jade, especially the clear green varieties, works on the physical and mental bodies, soothing nervous energy, grounding and calming the wearer. It will not absorb or hold any imbalances. It can impart great clarity to the re-

ceptive; centering the mind, creating an environment conducive to meditation and clear perception of Truth.

Jade will also remove any blocks in the wearer which might prevent the free flow of healing energy, allowing for the development of healing powers.

Jade and clear quartz, when used together, foster qualities of mental and physical strength, self-assuredness, wisdom, and healing energy.

JASPER

Jasper is found in many colors, among them browns, reds, greens, oranges, and greys. This mineral works best when set in silver. Any shape appealing to the wearer will be effective.

Jasper is one of the most grounding stones, serving to reconnect the wearer with the Earth, and then to strengthen that connection. It enhances the life force in man, repairing any tears in the aura, especially those caused by surgery or other physical trauma.

Jasper, when mounted with clear quartz crystal, has a calming, stabilizing effect on the wearer, imparting assuredness, strength, and vitality.

KUNDZITE

Kundzite is a pink to violet, clear gem, best faceted

and cut into an oval shape, mounted in white or yellow gold.

It works mainly on the spiritual and emotional bodies, calming and healing any turbulence or inappropriateness in these areas, clarifying the Truths to be learned through these centers. It then helps the wearer to incorporate this Wisdom on all levels.

When worn in combination with clear quartz crystal this stone has the magical effect of imparting a "peace that passeth all understanding" to the wearer through its gift of spiritual Wisdom on an emotional level.

LAPIS LAZULI/SODALITE

Lapis and sodalite are in the same mineral family, and although lapis is slightly more effective, they both have the same vibrational qualities. These opaque stones are blue, and the deeper the blue, the clearer the vibrational transmission. They can be any shape, but triangular cuts are especially effective. White or yellow gold, or silver can be used as a mounting medium.

Lapis and sodalite are excellent tools to remove blocks and to open up the third eye...the seat of higher wisdom and knowledge, the Eye of God. They are enhancers of the meditative state, clarifying and purifying energy received through the higher centers.

When combined with quartz crystal, these mystical stones encourage clarity of vision, higher Divine Wisdom, clairvoyance, and conscious awareness of the psychic powers inherent in man.

MALACHITE

Malachite is an opaque, vibrant green mineral frequently containing striations of a darker color. Any shape appealing to the wearer, mounted in silver, is most effective.

This gem is a very strong healer of the physical body in general; the heart area in particular. It draws any impurities out of the physical, emotional, and etheric bodies, and helps to repair any tears or weak places in the aura.

It is also an excellent healer of the eyes.

Malachite forms a bridge between the inner self and guides on an astral level, enhancing the intuitive skills of the wearer.

Using malachite in combination with clear quartz gives one emotional and physical strength, vibrant health, and a conscious link to the astral level of existence.

MOONSTONE

Moonstone is an opaque, pearly white mineral with a sheen which sometimes will form a star shape. It is best cut into a rounded, domed shape, set in either gold or silver.

This gem works on the astral and emotional bodies, sharpening intuitive skills, increasing awareness on a astral level. It has a Lunar energy which nourishes psychic abilities in the wearer.

When worn in combination with clear quartz crystal, it puts one in touch with the mystical side of life; the reality beyond the illusion. It should not be used by anyone who is not firmly grounded on a physical level. It is also very effective when used with amethyst, or rose quartz.

ONYX

Onyx can be almost any color, including red, grey, and even banded, but black is by far the most powerful, especially when cut into a smooth rectangular, squared, or oval shape, mounted in silver.

Onyx serves as a protector on a mystical level, preventing any drainage of energy by another. It stabilizes the emotions and strengthens the will on a physical as well as mental level. It will absorb all inappropriate energy directed toward the wearer; yet it will not store this energy, instead transforming it back to its original state of balance. Onyx is an excellent tool to use in psychometry, (the ability to know a person by holding an object belong-

ing to them). It also serves as a stimulant to the intuitive, psychic powers of the wearer.

When combined with clear quartz crystal, black onyx serves as a powerful protector, stabilizer, and enhancer of psychic energies and intuitive awareness.

OPAL

Opals can be green, white, black, red, violet, or blue. They all contain brilliant flashes of multi-colored fire within their depths. They are opaque stones and are best when cut into rounded shapes, mounted in gold.

Opals serve as a door into the inner world of spiritual awareness, through which man can enter into his mystical nature. They will increase the wearer's capacity for imaginative visualization, as well as the ability to form clear images on an astral level.

They have the ability to shatter illusions of reality and so can be very uncomfortable at times, and should not be used by those in need of a grounding influence. However, they are an excellent tool for one who desires to become an explorer of man's greater potential as a more fluid, spiritual being.

Because of the trapped liquids and gases within these stones—which cause the flashing lights—opals are very sensitive to any rapid change in the environment around them, and have been known to shatter or crack because of this.

When combined with clear quartz crystal, these gems foster creativity, astral awareness, and the fire of spiritual wisdom in the wearer.

ROSE QUARTZ

Rose quartz is a translucent to opaque, light to dark pink member of the quartz family. It is most effective when cut into rounded shapes, mounted in silver or gold.

This rose-colored mineral carries with it the quality of all accepting, all forgiving, Universal Love. It bathes the wearer in a feeling of softness, openness, and loving wonder at the miracle of existence. It works mainly on an emotional level through the heart center, healing any blocks that may exist.

When worn in combination with clear quartz crystal, it encourages a calm, loving, joyous, accepting attitude in the wearer; a feeling of at-oneness with all the Universe.

RUBY

Rubies are found in varying shades of red and pink, the clearer the stone and deeper the color, the better.

They are best cut into rounded shapes, faceted and mounted in gold.

This gem works on an emotional and spiritual level, nurturing the higher aspects of love and devotion to the Divine in everything. It creates a longing for Truth in the wearer and aids in the internalization of the fact that the Divine nature he recognizes in others is also found in himself.

Indeed, the ruby encourages a spiritual communion between the Divine in all things and the wearer, so that he becomes consciously aware of this exchange of loving energy.

The ruby, in combination with clear quartz, helps the wearer to become more open and giving, serving as a conduit for Divine Love, taking it in and then sending it out again to the whole Universe. In this way, it increases the strength of the physical, emotional and spiritual bodies, removing all blocks to create a free flow of God's Energy.

SAPPHIRE

Sapphires are found in blue, black, yellow, pink, white, and grey. They can be clear, translucent, or opaque, containing a sheen in the shape of a star.

The clear stones are best when cut in faceted, rounded shapes; the star sapphires, unfaceted and dome-shaped. Shades of blue seem to be most effective. They are best mounted in white-gold or silver.

Sapphires heal on a mental and spiritual level, cleansing and purifying these bodies in man. They give to man an assuredness of his spiritual nature, and the mental strength to relate to and act on this spirituality.

The star sapphires are excellent focal points for meditative states.

When combined with clear quartz crystals, they transmit qualities of mental and spiritual wisdom and purity of intent to the wearer.

TIGERS-EYE

Tigers-eye is a shiny, striated, opaque mineral found in shades of gold, brown, and blue. Shape doesn't seem to matter; however, they are best mounted in silver, or white gold.

This stone is a potent manifestor on the physical plane, easing the birth of ideas into reality. It also helps to work out any past-life connections or karmic ties by aiding the wearer in the recognition of these connections to others in this life. It helps in the knowing of what is important and necessary to this life, and what isn't, and then gives the strength to act on this knowledge, becoming master of one's desires.

Tigers-eye, when worn with clear quartz, encourages direct action, certainty of purpose, and strength of will.

TOPAZ

The topaz is found in several colors—blues, yellows, browns, pinks—but the blues and the golden hues seem to have the greatest effect on man.

Gems that are clear and faceted, mounted in white or yellow gold, are best. Either rounded or squared shapes are good to use.

The blue topaz is a stimulant to manifesting creatively on the physical plane. It aids in bringing creative energy from an astral, or spiritual, level into a physical expression, where it can be shared with all mankind.

Wearing blue topaz in combination with quartz crystal encourages the manifestation of spiritual awareness and creativity through the mental body into the physical plane.

Golden topaz serves as a stimulant, strengthening and expanding a sense of physical and emotional well-being. It is a good de-stressor, drawing any inappropriateness away from the aura, and protecting against depression.

It encourages a positive, open, joyful attitude when worn in combination with quartz crystal, adding a magnetic, loving dimension to the personality of the wearer. It also has somewhat of a grounding effect, its gift a feeling of being fully at home in the body, strong and assured of one's place in the world.

TURQUOISE

Turquoise is an opaque mineral found in all shades of green and blue. Any shape is effective, but they are best when mounted in silver.

This stone contains all the power of the Earth's energies, and will transfer this power and knowledge of the ways of the Earth to the wearer. It works on the physical and emotional bodies to give peace, serenity, and protection to those attracted to it.

It balances the physical and emotional bodies, keeping man's feet planted assuredly on the ground so that he can be free to experience the mysteries of life, and then relate them to the rest of his existence.

These stones also serve as a sort of shock absorber, taking in any trauma on an etheric level, before it can harm the wearer. There is a very loving energy surrounding it, connecting the heart with the mind.

Wearing turquoise in combination with clear quartz offers the wearer strength, a sense of purpose, creativity, and wisdom.

POSTSCRIPT

Remember, in the end, you are your best teacher. I have attempted in this book to give you a few guide-posts…a jumping-off place. Now it is up to you to learn how to swim. Just remember, you already know it all. Trust your intuition!!!

I also hope you will remember that anything you can do with the crystals, you can also do without them; the energy they are channeling is the same as the energy that is available to you in overwhelming abundance, once you have removed all resistance to accepting it.

The crystals are simply a tool to help you get out of your own way, so you can step joyously into your Divinity!

ALSO AVAILABLE FROM THE AUTHOR

NATURAL QUARTZ CRYSTALS

Selected with love and care for their metaphysical properties

Write us about yourself and what path you wish to take with the crystals (include a photo if you so desire). We will then select a crystal that is in tune with you and your purpose. Please specify clearly what type crystal you want and send a check or money order to include additional shipping and handling costs.

MEDITATION AND HEALING CRYSTALS ...... $10 - $200
ALTAR CRYSTALS .... $80 - $500
QUARTZ PENDANTS mounted in silver $25
silver and gemstones $35*
gold and amethyst or garnet $65
**malachite, bloodstone, carnelian, onyx etc.*
SERENITY VOLUME I
Beautiful original music created by Serenity on cassette tape. $10

shipping and handling charges:

TAPES $1
PENDANTS AND SMALL CRYSTALS $2.75
ALTAR CRYSTALS $5

Send orders to: **SERENITY PETERSON**
P.O. Box 41173
Nashville, Tenn. 37204

Please allow two to six weeks for delivery
(Tennessee residents must include sales tax)